Beautiful Birds

FLAMINGOS

By Bray Jacobson

Gareth Stevens
PUBLISHING

Please visit our website, www.garethstevens.com. For a free color catalog of all our high-quality books, call toll free 1-800-542-2595 or fax 1-877-542-2596.

Library of Congress Cataloging-in-Publication Data

Names: Jacobson, Bray, author.
Title: Flamingos / Bray Jacobson.
Description: Buffalo, New York : Gareth Stevens Publishing, [2024] |
Series: Beautiful birds | Includes index. |
Identifiers: LCCN 2023031201 | ISBN 9781538292501 (library binding) | ISBN 9781538292495 (paperback) | ISBN 9781538292518 (ebook)
Subjects: LCSH: Flamingos–Juvenile literature.
Classification: LCC QL696.C56 J34 2024 | DDC 598.3/5–dc23/eng/20230710
LC record available at https://lccn.loc.gov/2023031201

First Edition

Published in 2024 by
Gareth Stevens Publishing
2544 Clinton St.
Buffalo, NY 14224

Editor: Kristen Nelson
Designer: Andrea Davison-Bartolotta

Photo credits: Cover Evannovostro/Shutterstock.com; p. 1 Rosamar/Shutterstock.com; p. 5 Anhilator/Shutterstock.com; p. 7 Diegomezr/Shutterstock.com; pp. 9, 24 (bill) gary718/Shutterstock.com; p. 11 Jeff McGraw/Shutterstock.com; pp. 13, 24 (shrimp) Napat/Shutterstock.com; p. 15 Pieter JF Smit/Shutterstock.com; pp. 17, 24 (nest) SunflowerMomma/Shutterstock.com; p. 19 Jiri Balek/Shutterstock.com; pp. 21, 24 (chick) Ondrej Chvatal/Shutterstock.com; p. 23 Sergey Uryadnikov/Shutterstock.com.

Printed in the United States of America

CPSIA compliance information: Batch #CW24GS: For further information contact Gareth Stevens, New York, New York at 1-800-542-2595.

Contents

Look at that big,
pink bird!
It is a flamingo.

It has long, thin legs.
It has webbed feet.

It has a curved bill.
Its neck looks like an S.

Its feathers are pink! The color comes from its food.

It eats shrimp.
It eats snails.

It lives in a big group.
They fly together.

Nests are made of mud. They have a hole in the top.

Mothers lay one egg.
Both parents sit on it.

Babies are white!
They are called chicks.

Young turn pink
in a few years.

Words to Know

bill

chick

nest

shrimp

Index